Unleashing the Power: The Electrifying History of Tesla Corporation

Introduction

Discover the electrifying story behind one of the most innovative and groundbreaking companies of our time. In 'Unleashing the Power: The Electrifying History of Tesla Corporation,' delve into the captivating journey of Tesla, a company that has revolutionized the automotive industry and changed the way we perceive electric vehicles. From its humble beginnings to its current position as a global leader in sustainable energy solutions, this book takes you on a thrilling ride through the highs and lows, the triumphs and challenges, and the visionary genius that has shaped Tesla into an icon of innovation.

Prepare to be captivated as you uncover the untold secrets, the moments of brilliance, and the relentless pursuit of a sustainable future that define Tesla's remarkable history. With exclusive insights, behind-the-scenes stories, and interviews with key players, this book offers an intimate look into the birth, growth, and impact of the Tesla Corporation.

Whether you're a fan of electric vehicles, a technology enthusiast, or simply someone intrigued by the captivating journey of a company that has reshaped the automotive landscape, 'Unleashing the Power' will leave you inspired, awestruck, and yearning for the future. Join us as we explore the electrifying history of Tesla Corporation and uncover the untold stories that have propelled this visionary company to greatness.

Table of Contents

Chapter 1: Sparks of Innovation: The Birth of Tesla

Tesla Corporation was born out of a fiery determination to challenge the status quo and pave the way for a brighter, cleaner future. It all began with the convergence of brilliant minds and a shared vision to revolutionize the automotive industry. At the heart of this revolution stood Martin Eberhard, a man driven by an unwavering belief in the potential of electric vehicles.

As a young boy, Martin's fascination with science and technology ignited a fire within him. He would spend countless hours tinkering with gadgets and exploring the inner workings of machines. But it was his encounter with an electric vehicle that truly sparked his imagination and set him on a path of innovation.

The first time Martin laid eyes on an electric car, it was as if a bolt of lightning had struck him. The smooth, silent power of the vehicle left an indelible mark on his soul. He knew in that instant that electric transportation held the key to a future free from the shackles of fossil fuels. Fuelled by passion and a deep sense of responsibility to the environment, Martin set out on a mission to bring electric vehicles into the mainstream.

However, he soon realized that he couldn't achieve this grand vision alone. Enter Marc Tarpenning, a fellow entrepreneur with a passion for sustainable energy solutions. Marc shared Martin's belief in the potential of electric vehicles and saw the opportunity to create a company that could redefine the automotive landscape.

The union of Martin and Marc was nothing short of electric. Their shared enthusiasm, technical expertise, and complementary skills formed the foundation of what would become Tesla Corporation. Together, they set out to assemble a team of like-minded individuals who shared their unwavering commitment to building the world's finest electric cars.

The early days of Tesla were a whirlwind of excitement, challenges, and sleepless nights. The team poured their hearts and souls into developing a groundbreaking electric vehicle prototype that would captivate the world. Their dedication was unwavering, even in the face of skepticism and naysayers who dismissed their audacious goals.

It was during this pivotal time that Elon Musk, a visionary entrepreneur and relentless dreamer, entered the scene. Elon recognized the tremendous potential of Tesla and the electric vehicle movement. His unwavering belief in the power of technology to shape the future aligned perfectly with the mission of Tesla.

With Elon's strategic vision and leadership, Tesla Corporation received the injection of capital and expertise it needed to propel forward. Under his guidance, the company pushed boundaries and shattered expectations, introducing the world to the groundbreaking Tesla Roadster.

The Roadster was a revelation — an embodiment of Tesla's unwavering commitment to performance, sustainability, and design. With its sleek lines, blistering speed, and zero emissions, the Roadster captured the hearts of enthusiasts and skeptics alike. It was a testament to the collective genius of a team that refused to settle for mediocrity.

The success of the Roadster paved the way for Tesla's future endeavors. It showcased the immense potential of electric vehicles, dispelling doubts and inspiring a new wave of innovation. Tesla Corporation had set a new standard for excellence, and the world was starting to take notice.

In the next chapter, we will dive deeper into the groundbreaking technologies that propelled Tesla forward, explore the challenges they faced, and witness the pivotal moments that solidified Tesla's position as a trailblazer in the automotive industry.

Chapter 2: Driving Change: Revolutionizing the Automotive Industry

With the success of the Tesla Roadster, the world was awakened to the undeniable potential of electric vehicles. Tesla Corporation had shattered the perception that electric cars were mere novelties, proving that they could surpass the performance of traditional gasoline-powered vehicles. But Tesla's mission was far from complete. They set their sights on creating an electric vehicle that would not only captivate the hearts of enthusiasts but also appeal to a broader audience.

Thus, the Model S was born—a sleek, luxurious, and fully electric sedan that would revolutionize the automotive industry. It was a manifestation of Tesla's unwavering commitment to craftsmanship, innovation, and sustainability. The Model S quickly became an emblem of status and forward-thinking, attracting attention from drivers yearning for an alternative to the conventional.

The emotional journey of those who embraced the Model S was transformative. For many, it was more than a car; it represented a tangible step towards a greener future. The moment they slipped behind the wheel, the silent hum of the electric motor whispered promises of a cleaner, more sustainable world. Every press of the accelerator pedal ignited a surge of adrenaline, reaffirming their decision to be part of something greater than themselves.

Tesla's impact extended beyond the realm of individual vehicles. The company recognized the need for a robust charging infrastructure to alleviate concerns about range anxiety and ensure seamless long-distance travel. This led to the development of the Supercharger network — an ever-expanding web of high-speed charging stations strategically placed along popular travel routes.

The emotional resonance of the Supercharger network cannot be overstated. It symbolized freedom from the limitations of traditional refueling infrastructure. Drivers embarked on road trips with renewed confidence, knowing that they could recharge their vehicles quickly and conveniently, all while reducing their carbon footprint. The Supercharger network became a lifeline, connecting Tesla owners on a shared journey towards a sustainable future.

Tesla's commitment to sustainability extended beyond the vehicles themselves. Recognizing the pressing need to reduce dependence on fossil fuels, Tesla delved into the realm of energy storage. They introduced Powerwall, a sleek and compact home battery system that revolutionized the way we harness and consume energy. Powerwall empowered homeowners to store excess energy generated by renewable sources, such as solar panels, for use during times of high demand or power outages.

For those who embraced Powerwall, it was a transformative experience. They reclaimed control over their energy consumption, reducing reliance on the grid and embracing a more sustainable lifestyle. Each time the Powerwall kicked in during an outage, providing uninterrupted power, a surge of gratitude coursed through their veins. It was a tangible reminder that they were part of a movement, a movement towards a world powered by clean energy.

As Tesla continued to push boundaries, they faced numerous challenges. Skepticism, financial obstacles, and the relentless pursuit of perfection threatened to derail their progress. Yet, their unwavering dedication and the emotional investment of their supporters propelled them forward.

In the next chapter, we will explore Tesla's expansion beyond the road, their foray into energy solutions for the grid, and the audacious vision that has positioned Tesla as more than just an automotive company. Join us as we witness the emotional journey of those who embraced Tesla's mission and continue to drive change on a global scale.

Chapter 3: Electrifying the Road: The Rise of Electric Vehicles

The rise of electric vehicles marked a turning point in the history of transportation—a momentous shift towards a greener, more sustainable future. Tesla Corporation stood at the forefront of this electrifying revolution, championing the cause and transforming the perception of electric vehicles from niche novelties to mainstream marvels.

At the heart of Tesla's mission was the belief that electric vehicles should not be exclusive to the affluent but accessible to all. The Model 3 embodied this vision—a more affordable, mass-market electric car that aimed to revolutionize the way people thought about transportation. Its unveiling sent shockwaves through the industry, captivating the world with its sleek design, cutting-edge technology, and unprecedented demand.

For those who eagerly reserved a Model 3, the emotional journey was filled with anticipation and a profound sense of purpose. Each update from Tesla regarding production milestones and delivery timelines fueled their excitement, as they eagerly awaited the day they would finally take the wheel of their own electric dream. It was a transformative experience, not just in terms of the vehicle they were about to drive but also in the values they were embracing—a commitment to sustainability, innovation, and a future powered by clean energy.

As the Model 3 proliferated the roads, it became a symbol of hope—a tangible manifestation of the collective effort to reduce greenhouse gas emissions and combat climate change. Every time an individual passed a gas station without stopping, they felt a surge of pride, knowing that they were contributing to a cleaner, healthier planet. The emotional connection between owners of Tesla vehicles grew stronger, forming a community united by a shared passion for sustainability and a relentless pursuit of progress.

Tesla's impact extended far beyond individual owners. As more electric vehicles hit the streets, the air became cleaner, and the once-ubiquitous smog began to dissipate. Children could breathe deeply, free from the burden of polluted air. The beauty of nature became more vibrant, as the threat of exhaust fumes diminished. Electric vehicles were not just modes of transportation; they were a lifeline for a planet in peril.

The emotional resonance of Tesla's mission extended to the broader automotive industry. Traditional automakers, once dismissive of electric vehicles, started to take notice. They felt the groundswell of public demand, the emotional connection people had with their Teslas. Suddenly, the idea of electric vehicles no longer seemed like a distant dream, but an inevitable reality. Competitors emerged, vying for a piece of the electric vehicle market, spurred on by the emotional and environmental impact Tesla had made.

Tesla's audacious mission to accelerate the world's transition to sustainable energy was not without its challenges. The company faced logistical hurdles, supply chain constraints, and the relentless pressure to meet the ever-increasing demand for their vehicles. But amidst these trials, the emotional investment of their supporters remained steadfast. They believed in Tesla's vision and were willing to weather the storms, knowing that the rewards were not just personal, but planetary.

In the next chapter, we will explore Tesla's expansion into energy solutions, their groundbreaking work on energy storage, and the profound impact these innovations had on the grid and beyond. Join us as we witness the emotional journey of individuals who embraced Tesla's electric revolution and became catalysts for change in a world hungry for sustainable solutions.

Chapter 4: The Roadster Revolution: Unleashing the Power of Speed

The Tesla Roadster was more than just an electric vehicle; it was a revolution on wheels. With its sleek design, groundbreaking technology, and unparalleled speed, the Roadster shattered the limitations of electric transportation, captivating the world and redefining what was possible.

For those fortunate enough to witness the birth of the Roadster, the emotional journey was one of astonishment and wonder. As the curtains were drawn back, revealing the sleek lines and aerodynamic contours of the Roadster, hearts raced in anticipation. The room pulsed with a tangible energy, a collective gasp filling the air as the realization set in — that this was not just an electric car; it was a game-changer.

The Roadster possessed an otherworldly ability to accelerate from zero to sixty miles per hour in mere seconds. The G-forces pressed drivers into their seats as they were propelled forward, a surge of adrenaline coursing through their veins. It was an emotional experience — an exhilarating fusion of power, speed, and the knowledge that they were at the forefront of a seismic shift in automotive history.

For those who had doubted the capabilities of electric vehicles, the Roadster was a resounding answer — an emphatic declaration that electric cars could surpass even the most revered supercars in terms of performance. It left skeptics in awe, their preconceived notions shattered, and their hearts open to the electrifying possibilities that lay ahead.

The emotional resonance of the Roadster Revolution extended beyond the thrill of speed. It represented a triumph over the skeptics, a vindication for those who had championed the cause of electric transportation. It was a symbol of defiance — an audacious statement that Tesla was not content to simply compete; they were here to dominate.

As the Roadster roared down the streets, it left a trail of wonder and admiration in its wake. Enthusiasts and onlookers marveled at its silent power, the absence of engine noise replaced by an electric hum that whispered promises of a cleaner, more sustainable future. The emotional connection between driver and machine grew stronger with each passing mile, as they reveled in the knowledge that they were part of a revolution — a revolution that would forever change the automotive landscape.

The Roadster Revolution inspired a new generation of dreamers and visionaries, who dared to push the boundaries of what was possible. It ignited a passion for innovation, propelling engineers, designers, and enthusiasts to unlock the full potential of electric vehicles. It was no longer a question of "if" electric vehicles would dominate the roads, but rather "when" they would become the norm.

In the next chapter, we will explore Tesla's groundbreaking work on energy solutions, their relentless pursuit of sustainable power, and the emotional journey of those who harnessed the transformative potential of Tesla's innovations. Join us as we witness the impact of Tesla's energy solutions on the world and the emotional connection individuals formed with these revolutionary technologies.

Chapter 5: Powering the Future: Tesla's Energy Solutions

Tesla's mission to revolutionize transportation extended far beyond the realm of electric vehicles. They recognized that the key to a sustainable future lay in harnessing the power of renewable energy and transforming the way we consume and store electricity. With this in mind, Tesla set out to develop energy solutions that would empower individuals and communities to take control of their energy needs.

At the forefront of Tesla's energy revolution stood the Powerwall—a sleek and compact home battery system that captured the imagination of homeowners around the world. For those who embraced the Powerwall, it represented a leap towards energy independence and resilience. Each installation was more than just a technological upgrade; it was a declaration of self-sufficiency, a tangible step towards reducing reliance on the grid and embracing a cleaner, more sustainable lifestyle.

The emotional journey of those who installed a Powerwall was transformative. They felt a newfound sense of empowerment, knowing that they had the ability to store excess energy generated by renewable sources, such as solar panels, for use during times of high demand or power outages. Each time the Powerwall kicked in, seamlessly providing uninterrupted power, a surge of gratitude coursed through their veins. It was a testament to their commitment to sustainability and a tangible reminder that they were actively shaping a future powered by clean energy.

Beyond individual homes, Tesla ventured into the realm of commercial and utility-scale energy solutions. They introduced the Powerpack—a larger-scale battery system capable of storing vast amounts of energy generated by renewable sources. Powerpack installations revolutionized the way communities and businesses approached energy consumption, enabling a smoother integration of renewable energy into the grid and paving the way for a more sustainable future.

The emotional impact of Tesla's energy solutions extended beyond the individual users. As more homes and businesses adopted these technologies, the collective impact reverberated through communities. Neighborhoods transformed into microgrids, interconnected and resilient. The reliance on fossil fuels diminished, replaced by a network of clean energy sources and energy storage systems. It was a testament to the power of collective action—a reminder that small choices made by individuals could have a profound impact on a global scale.

Tesla's vision extended to the realm of solar energy as well. They introduced solar roof tiles that seamlessly blended renewable energy generation with aesthetics, transforming the very fabric of our built environment. The emotional connection individuals formed with these solar roofs was profound. Each time they looked up at their homes, they saw more than just a roof; they saw a statement—a commitment to a sustainable future and a visual reminder of their contribution to combating climate change.

The emotional resonance of Tesla's energy solutions went beyond the technology itself. It was a realization that the power to shape a sustainable future was in our hands — that by embracing these innovations, we could collectively reduce our carbon footprint and create a world powered by clean energy. It awakened a sense of hope, inspiring individuals to take action and become agents of change in their own communities.

In the next chapter, we will explore Tesla's audacious vision for a network of Gigafactories, their commitment to scaling up production, and the emotional journey of those who witnessed the transformation of Tesla into a global powerhouse. Join us as we witness the impact of these manufacturing giants and the emotional connection individuals formed with the idea of mass-scale sustainable production.

Chapter 6: Breaking Barriers: Supercharging the Electric Vehicle Network

As Tesla set out to revolutionize the automotive industry, they faced a significant challenge: the limited infrastructure for electric vehicles. Traditional charging stations were scarce, and range anxiety cast a shadow of doubt on the minds of potential electric vehicle owners. Tesla knew that to truly unleash the potential of electric transportation, they had to break through these barriers.

And so, the Supercharger network was born—an audacious vision to create a global network of high-speed charging stations strategically placed along popular travel routes. It was a game-changer, a bold declaration that long-distance travel in electric vehicles was not only possible but also convenient and efficient.

For those who embarked on cross-country journeys with their Tesla vehicles, the emotional journey was one of freedom and liberation. They bid farewell to the limitations of traditional refueling infrastructure, knowing that they had access to a network that would enable them to recharge their vehicles quickly and conveniently. Each Supercharger stop became an opportunity for connection—a chance to meet fellow Tesla enthusiasts, exchange stories, and share the excitement of being part of a global movement.

The emotional resonance of the Supercharger network extended beyond individual journeys. It was a symbol of progress — a tangible manifestation of Tesla's commitment to building the necessary infrastructure to accelerate the adoption of electric vehicles. The expansion of the Supercharger network was met with anticipation and celebration, as communities around the world welcomed these charging stations with open arms. It was a testament to the power of connectivity, of coming together to build a future where sustainable transportation was accessible to all.

The Supercharger network also symbolized Tesla's dedication to customer experience. Each charging station was meticulously designed to provide a seamless and enjoyable experience. Charging stalls were equipped with amenities, allowing drivers to relax, work, or grab a bite to eat while their vehicles recharged. The emotional impact of these thoughtful touches cannot be overstated — it transformed the perception of charging from a chore to a moment of respite, a chance to recharge not only their vehicles but also their own energy and well-being.

Tesla's relentless pursuit of expanding the Supercharger network faced its fair share of challenges. They had to navigate regulatory hurdles, overcome logistical complexities, and keep up with the exponential growth of their vehicle fleet. Yet, their emotional investment in providing a superior charging experience never wavered. Each new Supercharger installation brought with it a sense of achievement, a validation of their vision, and a renewed commitment to transforming the transportation landscape.

In the next chapter, we will explore Tesla's expansion beyond the road, their ventures into energy storage, and the emotional journey of individuals who embraced these innovative solutions. Join us as we witness the impact of Tesla's energy storage solutions on the grid and the emotional connection individuals formed with these transformative technologies.

Chapter 7: Beyond the Road: Exploring Tesla's Expansion

Tesla Corporation's vision extended far beyond the confines of the road. They recognized that the transformation of transportation was just the beginning—a stepping stone towards a future powered by sustainable energy. With this in mind, Tesla embarked on a journey of exploration, pushing the boundaries of innovation and reshaping industries beyond automotive.

One such venture was energy storage—a realm where Tesla's technological prowess truly shone. They introduced the Powerpack, a larger-scale battery system capable of storing vast amounts of energy. It revolutionized the way communities, businesses, and even entire regions approached energy consumption. The emotional impact of Tesla's energy storage solutions was profound.

For those who witnessed the implementation of Powerpack installations, it was a moment of awe and inspiration. It signaled a paradigm shift in the way we harnessed and utilized energy—a departure from the limitations of traditional power grids and a move towards a more sustainable and resilient future. The emotional journey of individuals who embraced this technology was transformative. They felt a renewed sense of control over their energy consumption and a deeper connection to the planet they called home. Each time they tapped into the stored energy during peak demand or power outages, a surge of gratitude coursed through their veins. It was a reminder that they were part of a movement—a movement that aimed to reshape our relationship with energy and create a more sustainable world.

But Tesla's exploration did not stop at energy storage. They set their sights even further, reaching for the stars. Tesla, under the visionary leadership of Elon Musk, ventured into the realm of space exploration with SpaceX. The emotional impact of this audacious endeavor cannot be overstated. It ignited a sense of wonder and possibility — a belief that humanity's future extended beyond the confines of Earth.

For those who followed Tesla's journey into space, it was an emotional rollercoaster of anticipation, triumph, and heart-stopping moments. They watched in awe as rockets were launched, payloads delivered, and boundaries pushed. Each successful mission became a testament to the indomitable spirit of human exploration and a reminder that Tesla was not just an automotive company but a catalyst for progress across industries.

The emotional resonance of Tesla's expansion extended to other ventures as well. Their acquisition of SolarCity brought solar energy to the forefront, integrating renewable power generation seamlessly into the fabric of our built environment. The emotional connection individuals formed with solar energy was profound. Each time they looked at their solar panels, soaking in the sunlight and converting it into clean, sustainable energy, they felt a deep sense of pride and purpose. It was a visual reminder that they were actively contributing to a future powered by the sun.

In the next chapter, we will explore the dynamic partnership between Tesla and Elon Musk, their shared vision for innovation, and the emotional journey of those who witnessed their collaboration. Join us as we dive into the world of Tesla and Elon Musk, where dreams are turned into reality and a relentless pursuit of a better future is the driving force.

Chapter 8: Tesla and Elon Musk: The Dynamic Duo

Elon Musk — the name alone evokes a sense of awe, a symbol of visionary genius and relentless ambition. His partnership with Tesla Corporation has been nothing short of extraordinary — a dynamic duo that has propelled the company to unprecedented heights and reshaped industries in their wake.

Elon Musk's entry into Tesla's narrative was a catalyst for exponential growth and a game-changer in the truest sense of the word. With his strategic vision and unwavering belief in the potential of Tesla, he injected new life and boundless energy into the company. The emotional impact of Elon's presence was profound — it brought a renewed sense of purpose and ignited a fire within the Tesla team and supporters worldwide.

For those who witnessed the synergy between Tesla and Elon Musk, it was an emotional journey of inspiration and awe. Elon's unwavering commitment to a sustainable future and his audacious goals resonated deeply with the Tesla community. They felt a kinship — a shared belief that the status quo could be challenged, that barriers could be shattered, and that a better world was within reach.

The emotional connection between Tesla and Elon Musk extended beyond the boardroom. His visionary leadership and charismatic persona ignited a sense of hope and possibility, inspiring not only Tesla employees but also a generation of dreamers and entrepreneurs. Elon's relentless pursuit of innovation and his willingness to take risks pushed the boundaries of what was deemed possible, propelling the entire industry forward.

Yet, the emotional journey of Tesla and Elon Musk was not without its challenges. The weight of expectations, the scrutiny of the public eye, and the relentless pursuit of perfection took their toll. The world watched with bated breath as Tesla faced production hurdles, financial obstacles, and the pressure to deliver on their promises. But through it all, Elon's unwavering belief in the mission of Tesla and his emotional investment in its success never faltered. He stood at the helm, guiding the company through stormy seas with a steady hand and an unwavering vision.

The partnership between Tesla and Elon Musk was a testament to the power of collaboration and shared purpose. It was a reminder that when visionary minds come together, incredible things can be achieved. The emotional connection between Tesla and Elon Musk symbolized the triumph of imagination over skepticism, of boldness over complacency, and of unwavering belief over doubt.

In the next chapter, we will explore the iconic Model S, Tesla's pioneering luxury electric vehicle that captivated the world. Join us as we delve into the emotional journey of those who embraced the Model S and witnessed the birth of a new era of automotive excellence.

Chapter 9: The Model S: Pioneering Luxury Electric Vehicles

The Model S—a symbol of elegance, innovation, and unrivaled performance. Tesla's pioneering luxury electric vehicle shattered expectations, redefining what it meant to drive in style while leaving a lasting emotional impact on those fortunate enough to experience it.

From the moment one set eyes on the Model S, a surge of emotion swept over them. Its sleek lines and aerodynamic curves spoke of a harmonious blend of form and function. Every detail, meticulously crafted, exuded a sense of refined elegance. As drivers approached the vehicle, anticipation grew, knowing that they were about to embark on a journey that would forever change their perception of luxury cars.

Stepping inside the Model S was a revelation—an emotional experience that transcended traditional notions of automotive interiors. The absence of traditional dials and buttons was replaced by a minimalist, futuristic cockpit. The large touchscreen display, like a portal to the vehicle's soul, invited drivers to engage with the car in a way that felt both intuitive and immersive. It was a harmonious fusion of technology and design—a space that embraced the driver and cocooned them in a world of comfort and sophistication.

But the emotional journey of the Model S went far beyond its captivating aesthetics. It was the performance that truly took drivers' breath away. The instant torque, delivered silently and seamlessly, pressed them back into their seats as they accelerated with exhilarating speed. The raw power of electric propulsion combined with the refined grace of a luxury vehicle created a symphony of emotion—an experience that etched itself into the very core of their being.

Each time drivers took the Model S on the open road, they felt a sense of liberation — a freedom to explore without compromise. The anxiety of range limitations dissolved as they navigated effortlessly from one charging station to another, confident in the knowledge that they could embark on long-distance journeys without sacrificing comfort or style. The Model S became a trusted companion — a vehicle that not only fulfilled their transportation needs but also became an extension of their identity, a testament to their commitment to a sustainable future.

The emotional connection between drivers and the Model S extended beyond the vehicle itself. It was a bond formed with a community of like-minded individuals who shared a passion for innovation and sustainability. They reveled in the camaraderie and mutual support that arose from being part of the Tesla family. Each encounter with fellow Model S owners was an opportunity to exchange stories, share charging tips, and celebrate the transformative power of electric vehicles.

The impact of the Model S was far-reaching. It challenged the prevailing narrative that electric vehicles were mere novelties or compromises. It proved that luxury, performance, and sustainability could coexist harmoniously. The emotional resonance of the Model S inspired a shift in perception — a realization that the future of automotive excellence lay in the realm of electric power.

In the next chapter, we will explore the introduction of the Model X and Model 3, witnessing the expansion of Tesla's vehicle lineup and the emotional journey of those who embraced these groundbreaking models. Join us as we continue to unravel the story of Tesla's evolution and the profound impact they had on the automotive industry.

Chapter 10: The Model X and Model 3: Expanding the Electric Revolution

With the success of the Model S, Tesla set its sights on expanding its vehicle lineup, determined to bring the electric revolution to a wider audience. Enter the Model X—a family-friendly SUV that blended versatility, luxury, and electric performance into a captivating package. It was a testament to Tesla's commitment to innovation and their unwavering belief that electric vehicles could transcend traditional boundaries.

For those who embraced the Model X, the emotional journey was one of transformation—a realization that an electric vehicle could cater to their practical needs without compromising on style or functionality. The iconic falcon-wing doors, soaring skywards with grace, became a symbol of ingenuity and audacity. They opened up a world of possibilities, offering convenient access to the spacious interior and capturing the imagination of onlookers.

Inside the Model X, families discovered a sanctuary—a space that seamlessly blended comfort, technology, and safety. The expansive panoramic windshield provided an unobstructed view of the world, while the cutting-edge features and advanced driver-assistance systems created a sense of security and peace of mind. The emotional connection formed with the Model X extended beyond being just a mode of transportation—it became a trusted companion on adventures, a haven for cherished memories, and a testament to a commitment to sustainability and a better future for generations to come.

But Tesla's journey towards accessibility didn't stop there. They had a bolder vision — a mass-market electric vehicle that would disrupt the automotive landscape and accelerate the transition to sustainable transportation for all. The Model 3 was born — a revolutionary sedan that embodied Tesla's unwavering commitment to excellence and affordability.

The emotional impact of the Model 3 cannot be overstated. Its introduction marked a turning point — an invitation for a wider audience to embrace the electric revolution. As the reservations poured in, a wave of excitement swept across the globe. Individuals from all walks of life eagerly awaited their turn to experience a Tesla firsthand, to be part of a movement that aimed to reshape the future of transportation.

For those who became proud owners of the Model 3, the emotional journey was one of empowerment — a realization that sustainability and luxury were no longer reserved for the few, but accessible to many. Each time they slid behind the wheel, they felt a surge of pride and purpose, knowing that they were contributing to a sustainable future while experiencing the sheer joy of electric performance.

The Model 3 became more than just a vehicle — it became a symbol of hope and progress. It inspired a new generation of electric vehicle enthusiasts and shifted the perception of what an affordable mass-market car could be. The emotional connection formed with the Model 3 transcended the individual — it became a collective movement, a testament to the power of a shared vision and the belief that a better future was within reach.

In the next chapter, we will explore Tesla's commitment to sustainability and energy solutions, witnessing the emotional journey of those who embraced Tesla's innovations and harnessed the transformative power of sustainable energy. Join us as we continue to unravel the story of Tesla's impact and the emotional connection individuals formed with the company's mission to accelerate the world's transition to sustainable energy.

Chapter 11: Sustainability and Energy Solutions: Harnessing the Power of Transformation

Tesla's mission to accelerate the world's transition to sustainable energy extended beyond their groundbreaking electric vehicles. They recognized the urgent need to address the pressing challenges of climate change and sought to revolutionize the way we generate, store, and consume energy.

One of the most significant steps Tesla took towards sustainability was the acquisition of SolarCity — an industry leader in solar energy solutions. This marked a pivotal moment — a fusion of innovation and environmental consciousness. For those who embraced Tesla's solar power solutions, the emotional journey was one of empowerment and responsibility.

As solar panels adorned rooftops, harnessing the power of the sun, individuals felt a deep connection to the planet and a renewed sense of purpose. Each moment they watched their meters spin backward, offsetting their energy consumption with clean, renewable power, they were filled with a sense of pride and gratitude. It was a reminder that they were actively contributing to a more sustainable future, one where reliance on fossil fuels was diminished, and the potential for a cleaner world was realized.

Tesla's commitment to sustainability went beyond individual solar installations. They recognized the need for scalable solutions that could impact entire communities and regions. With this in mind, they ventured into the realm of energy storage. Tesla's Powerpack and Powerwall solutions revolutionized the way we store and consume energy, providing a seamless integration of renewable power into our daily lives.

For those who embraced Tesla's energy storage solutions, the emotional journey was one of freedom and self-reliance. The ability to store excess energy generated by renewable sources, such as solar panels, meant that individuals could take control of their energy consumption, reduce their reliance on the grid, and embrace a more sustainable lifestyle. Each time the stored energy powered their homes during peak demand or in the face of an outage, they felt a surge of gratitude and resilience. It was a tangible reminder that they were part of a transformative movement—a movement that harnessed the power of innovation to shape a greener, more sustainable world.

Beyond solar power and energy storage, Tesla embraced sustainability in their manufacturing processes. Their commitment to reducing their carbon footprint led to the establishment of Gigafactories—massive production facilities designed to streamline and scale up the manufacturing of electric vehicles, batteries, and other sustainable technologies.

For those who witnessed the birth of the Gigafactories, the emotional journey was one of hope and inspiration. They saw these manufacturing giants not just as buildings, but as beacons of progress — a testament to Tesla's dedication to a sustainable future. The Gigafactories represented more than just mass production; they symbolized a paradigm shift in the way we think about manufacturing — a shift towards clean, sustainable processes that would leave a lasting impact on future generations.

In the next chapter, we will explore Tesla's vision for a future powered by renewable energy, their audacious goals, and the emotional journey of those who embraced their vision. Join us as we continue to unravel the story of Tesla's impact and the profound connection individuals formed with the company's mission to reshape the world through sustainable energy solutions.

Chapter 12: A Vision for the Future: Tesla's Audacious Goals

Tesla has always been a company guided by a bold vision—a vision that extends far beyond the present and embraces the potential of tomorrow. With audacity and unwavering determination, they have set their sights on revolutionizing not only transportation but also the very fabric of our existence.

One of Tesla's most ambitious goals is the development of autonomous driving technology—an innovation that promises to redefine the way we navigate the world. For those who have witnessed the advancements in Tesla's Autopilot and Full Self-Driving capabilities, the emotional journey has been one of awe and anticipation.

The idea of a world where cars can navigate themselves, where roads are safer, and where commuting becomes a moment of relaxation or productivity, fills the heart with excitement. It is a vision that taps into the deepest recesses of our collective imagination, challenging the limits of what we thought was possible. The emotional connection formed with Tesla's autonomous driving technology transcends the mundane—it represents the realization of a dream, a leap into a future where innovation and convenience coexist harmoniously.

But Tesla's vision reaches even further, extending beyond the confines of Earth itself. Elon Musk, with his unyielding determination and insatiable curiosity, founded SpaceX—a company dedicated to the exploration and colonization of other planets. The emotional impact of SpaceX's audacious goals cannot be understated—it sparks a sense of wonder, a belief that the boundaries of human existence can be pushed beyond the stars.

For those who followed SpaceX's endeavors, the emotional journey has been one of inspiration and hope. Each rocket launch, each successful mission, reminds us that we are part of a species capable of extraordinary feats. It stirs a deep longing to explore the unknown, to reach for new frontiers, and to embrace a future where interplanetary travel is not just a work of science fiction, but a tangible reality.

The emotional connection between Tesla and its followers goes beyond the vehicles and the technologies they produce. It is a connection rooted in a shared belief—a belief that through innovation, determination, and an unwavering commitment to sustainability, we can create a future that transcends our wildest dreams.

In the next chapter, we will explore the emotional impact of Tesla's global influence, the community that has formed around the brand, and the stories of individuals who have been transformed by their experiences with Tesla. Join us as we delve into the profound connections that have been forged, and witness the power of a company that not only seeks to change the world but also touches the hearts and souls of those who encounter it.

Chapter 13: The Tesla Community: Connecting Hearts and Minds

Tesla is more than just a company — it is a movement, a community that extends far beyond the confines of the automotive industry. It is a gathering of like-minded individuals who share a common passion for innovation, sustainability, and the belief that a better future is within reach. Within this community, hearts and minds connect, forging bonds that transcend borders and backgrounds.

For enthusiasts and fans of Tesla, the emotional journey begins with a spark — an encounter that ignites a curiosity and fascination. It might be a Tesla vehicle gliding silently down the street or a news story that captures their attention. From that moment, a flame is kindled — a desire to learn more, to be part of something greater than themselves.

As these individuals dive deeper into the world of Tesla, they discover a vibrant and passionate community — a network of owners, enthusiasts, and advocates who share their love for the company and its mission. Online forums, social media groups, and local meetups become their virtual gathering places, where they can connect, share stories, and exchange insights. It is within these digital spaces that the emotional bond between Tesla enthusiasts grows stronger.

The Tesla community is a testament to the power of shared experiences. Members exchange tales of their first Tesla moments — the exhilaration of accelerating in silence, the joy of driving on clean energy, and the pride of being part of a movement that aims to reshape the world. These stories become threads that weave together a tapestry of emotion, forming a collective narrative of hope, inspiration, and the relentless pursuit of a better future.

The emotional impact of the Tesla community extends beyond the digital realm. Local Tesla clubs and events bring enthusiasts together in the physical world, providing an opportunity for face-to-face interactions and the forging of lifelong friendships. Whether it's a Tesla meetup at a charging station, a road trip organized by like-minded owners, or a gathering to celebrate the latest innovations, these events become catalysts for deep connections and shared moments of joy.

For Tesla owners, the emotional journey within the community is even more profound. Owning a Tesla is not merely a transaction — it is an initiation into a close-knit family. Each time they encounter another Tesla on the road, a wave of camaraderie washes over them. It is a silent acknowledgement, a recognition of their shared commitment to sustainability and innovation. The Tesla community becomes their extended family, a support system that celebrates milestones, offers guidance, and rallies around one another during challenging times.

Within the Tesla community, emotions run deep. From the excitement of receiving a new Tesla delivery to the pride of showcasing their vehicle at local events, each milestone is celebrated and amplified by the collective energy of the community. Members uplift one another, providing encouragement, inspiration, and a sense of belonging.

In the next chapter, we will explore the impact of Tesla's global influence, the stories of individuals who have been transformed by their experiences with the company, and the emotional connection between Tesla and its followers. Join us as we delve into the profound influence that Tesla has had on the hearts and minds of people around the world, and witness the power of a community united by a shared vision.

Chapter 14: The Impact of Tesla's Global Influence

Tesla's impact extends far beyond the confines of the automotive industry. It has ignited a global revolution — a movement towards sustainability, innovation, and a future powered by clean energy. The emotional resonance of Tesla's influence is palpable, as hearts and minds are touched, perceptions are transformed, and the world embraces a new vision for the future.

One of the most significant impacts of Tesla's global influence has been the shift in public perception. Once considered a niche market, electric vehicles have now become a symbol of progress, luxury, and environmental consciousness. Tesla's pioneering work has shattered the notion that electric cars are compromises, showing the world that they can surpass the performance, style, and convenience of traditional vehicles. The emotional journey of those who have witnessed this transformation is one of empowerment — an affirmation that individual choices can shape the collective consciousness and pave the way for a sustainable future.

Tesla's influence has extended beyond the automotive industry, inspiring change in energy production, manufacturing processes, and technological innovation. The company's commitment to sustainability and renewable energy has prompted other industries to reevaluate their practices and embrace more eco-friendly solutions. The emotional impact of this ripple effect is profound — it instills a sense of hope and possibility, as individuals and organizations recognize the power of collective action and the capacity to create a better world.

Tesla's influence has also sparked a wave of innovation, with other companies racing to develop electric vehicles and sustainable technologies. The emotional journey of those who have been touched by Tesla's influence is one of inspiration and drive. They witness the transformative power of one company's vision and are propelled to contribute to the sustainable movement in their own ways. It is a testament to the far-reaching impact of Tesla's influence — that it not only changes industries but also fuels a broader cultural shift towards a more sustainable and responsible mindset.

But perhaps the most profound impact of Tesla's global influence lies in the hearts and souls of individuals who have encountered the company. The emotional connection formed with Tesla's mission and vision is transformative — it awakens a sense of purpose, a desire to be part of something greater than oneself. It fuels a relentless pursuit of innovation, sustainability, and a better future. The emotional journey of those touched by Tesla's influence is one of personal growth, as they find solace, inspiration, and a deep sense of fulfillment in aligning their values with a company that is changing the world.

In the next chapter, we will explore the questions that most people have about Tesla, addressing common inquiries, clarifying misconceptions, and providing insights into the company's journey. Join us as we delve into the most frequently asked questions and provide meaningful answers, shedding light on the intricacies of Tesla's story and the impact it continues to have on the world.

Chapter 15: Unveiling the Answers: Addressing Common Questions About Tesla

1. "Is Tesla just an electric car company?"

No, Tesla is much more than that. While they are known for their groundbreaking electric vehicles, Tesla is also involved in energy storage, renewable energy solutions, and sustainability initiatives. Their vision extends beyond transportation, aiming to accelerate the world's transition to sustainable energy in all its forms.

2. "How does Tesla compare to traditional automakers?"

Tesla stands apart from traditional automakers in terms of their commitment to electric vehicles and sustainability. They have been at the forefront of the electric revolution, pioneering innovations in battery technology, autonomous driving, and charging infrastructure. Their focus on clean energy and cutting-edge design sets them apart from the competition.

3. "Are Tesla vehicles truly sustainable?"

Yes, Tesla vehicles are designed with sustainability in mind. They produce zero tailpipe emissions and have a significantly lower environmental impact compared to traditional gasoline-powered cars. Tesla's dedication to sustainability extends to their manufacturing processes, energy solutions, and the entire lifecycle of their vehicles.

4. "What is the Supercharger network, and how does it work?"

The Supercharger network is Tesla's global network of high-speed charging stations. It enables Tesla owners to charge their vehicles quickly and conveniently, especially during long-distance travel. The Supercharger stations are strategically placed along popular travel routes, providing a seamless charging experience for Tesla drivers.

5. "Can I use non-Tesla charging stations for my Tesla vehicle?"

While Tesla vehicles can be charged using non-Tesla charging stations with adapters, the Supercharger network provides the fastest and most convenient charging experience specifically designed for Tesla vehicles.

6. "What is Autopilot, and how does it work?"

Autopilot is Tesla's advanced driver-assistance system that utilizes a combination of sensors, cameras, and AI technology to enable semi-autonomous driving. It assists with tasks such as lane centering, adaptive cruise control, and automated parking. However, it is essential for drivers to remain attentive and in control of the vehicle at all times.

7. "Are Tesla vehicles safe?"

Tesla vehicles have earned top safety ratings from various organizations around the world. Their advanced safety features, including collision avoidance systems and robust structural design, prioritize the protection of both occupants and pedestrians.

8. "Why are Tesla vehicles considered luxury cars?"

Tesla vehicles are often considered luxury cars due to their cutting-edge technology, superior performance, and refined design. The attention to detail, luxurious interiors, and advanced features position Tesla as a high-end brand in the automotive industry.

9. "What is the future of Tesla?"

The future of Tesla holds exciting possibilities. With their ongoing commitment to innovation, sustainability, and expanding their vehicle lineup, Tesla aims to continue leading the way in electric transportation. They also have ambitious goals in autonomous driving, renewable energy solutions, and even space exploration with SpaceX.

10. "How can I become part of the Tesla community?"

Becoming part of the Tesla community is as simple as embracing Tesla's mission and values. Engage with other enthusiasts online, attend Tesla events and meetups, and consider owning or supporting Tesla products. The community welcomes those who share a passion for sustainability, innovation, and a better future.

Chapter 16: Tesla's Legacy: A Journey of Innovation and Inspiration

Tesla's legacy is one of innovation and inspiration—a testament to the power of human ingenuity and unwavering belief in a sustainable future. From humble beginnings to global recognition, Tesla has left an indelible mark on the world, transforming industries and touching the hearts of individuals who have encountered its mission.

The emotional resonance of Tesla's legacy is felt in every aspect of its journey. It is seen in the faces of individuals who have experienced the thrill of driving a Tesla—the smiles that spread across their faces as they accelerate with unparalleled electric performance. It is heard in the stories of those who have embraced sustainable energy solutions, witnessing the transformative power of harnessing the sun's energy and reducing their carbon footprint.

Tesla's legacy is also shaped by the emotional connection individuals have formed with the company's visionary leader, Elon Musk. His relentless pursuit of innovation, audacious goals, and unwavering commitment to a better future have inspired millions around the world. Elon's boldness and determination serve as a beacon of hope and a reminder that individuals can make a profound impact on the world.

The emotional journey of those who have encountered Tesla's legacy is one of empowerment—a realization that they too can contribute to a more sustainable world, that their choices matter, and that innovation knows no bounds. Tesla's legacy sparks a fire within their hearts—a fire that fuels their passion, their drive, and their belief in a future where sustainability and innovation coexist harmoniously.

Tesla's impact on the automotive industry is undeniable. They have shattered perceptions, disrupted traditional norms, and redefined what it means to drive a car. The emotional connection individuals have formed with Tesla's vehicles goes beyond transportation—it represents a lifestyle, a statement of values, and a commitment to a better future. Each time they sit behind the wheel, they are reminded that they are part of a movement—a movement that is transforming the way we interact with our environment and challenging the status quo.

But Tesla's legacy extends far beyond the automotive industry. Their work in energy storage, renewable energy solutions, and even space exploration has left an indelible mark on the world. Their influence has inspired other companies to innovate, governments to enact policies that promote sustainability, and individuals to rethink their impact on the planet.

The emotional journey of those touched by Tesla's legacy is one of hope and possibility. It is a belief that a better future is not only within reach but also achievable through the collective efforts of individuals, companies, and communities. Tesla has provided a blueprint—a shining example of what is possible when we dare to dream, when we embrace innovation, and when we unite in our pursuit of a sustainable world.

www.ingramcontent.com/pod-product-compliance
Lightning Source LLC
Chambersburg PA
CBHW072238230526
45466CB00025B/2103